The Human Brain

Cameron Macintosh

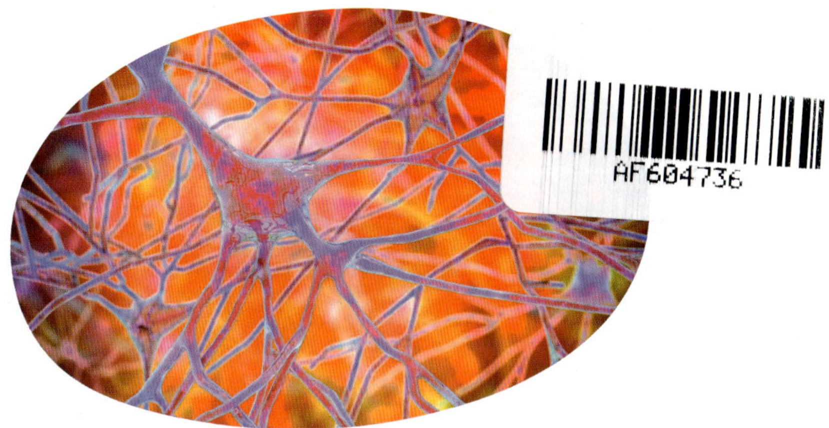

Contents

The Human Brain

What the Brain Does

The brain is one of the largest **organs** in the human body. It is a complex organ that controls a huge range of vital **functions** for every person. The brain allows people to observe and understand the world around them, and to interact with the world. It does this by gathering information through the five senses: sight, touch, hearing, smell and taste.

The brain allows us to gather information from our senses, such as our sense of smell.

When people consider the brain's functions, they often think of abilities such as thought and memory. The brain, however, is responsible for so much more than this. It controls many functions that people are unaware of, such as breathing, as well as physical **sensations**, like sensing temperature and feeling hunger. The brain also controls all the movements of the body.

The brain performs these functions partly through its role in the nervous system. This is a system which links the brain to the entire body through the **spinal cord**. An incredibly complex network of **nerves** spreads out from the spinal cord to every part of the body.

About three-quarters of the brain is made up of water. Drinking plenty of water helps the brain to function at its best.

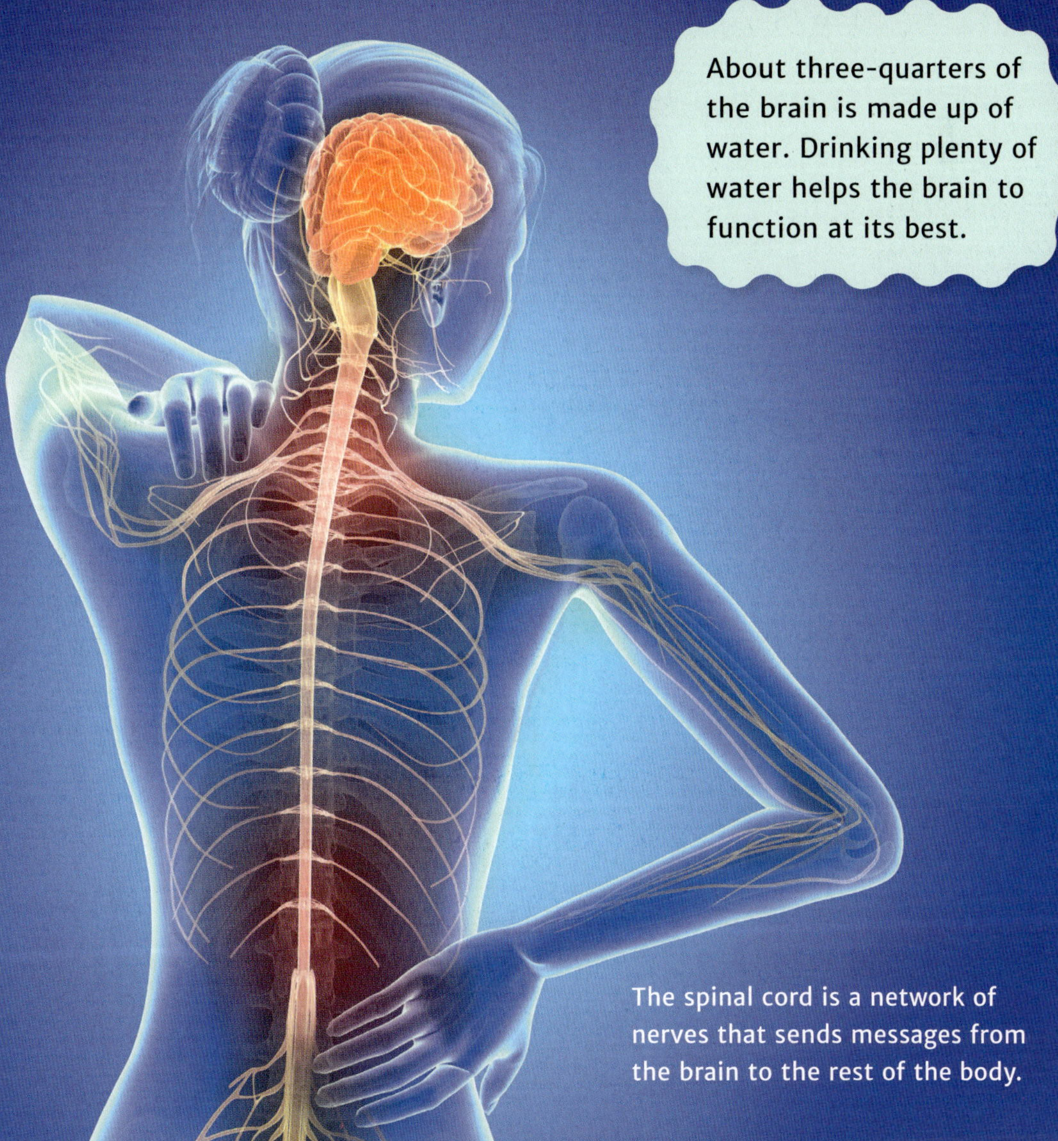

The spinal cord is a network of nerves that sends messages from the brain to the rest of the body.

Inside the Cranium

The brain is located in the top part of the skull, which is called the cranium. The cranium is made up of eight bones that are joined together by thick bands of **tissue** called sutures (pronounced *soo-chers*).
The cranium also contains special fluid and tissue that protects the brain. These are found between the brain and the inside of the skull.

The Cranial Bones

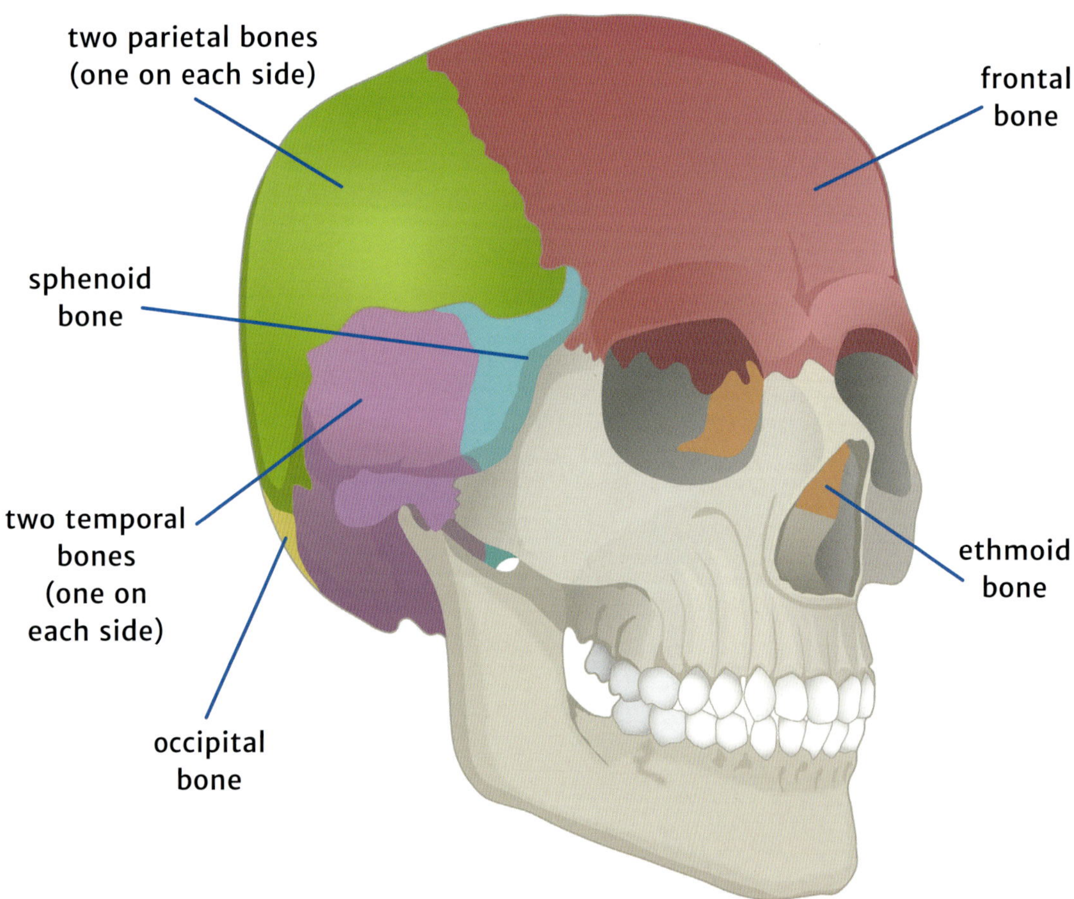

The brain itself has three main sections:

- the cerebrum (pronounced *suh-ree-brum*)
- the brainstem
- the cerebellum (pronounced *se-ruh-bell-um*)

The largest of these is the cerebrum.

These sections, particularly the cerebrum, are made up of smaller parts of different shapes and sizes. Each part performs specific functions, although the parts often work in cooperation with each other. Different parts of the brain can also contribute to the same function. For example, many parts of the brain contribute to people's ability to see.

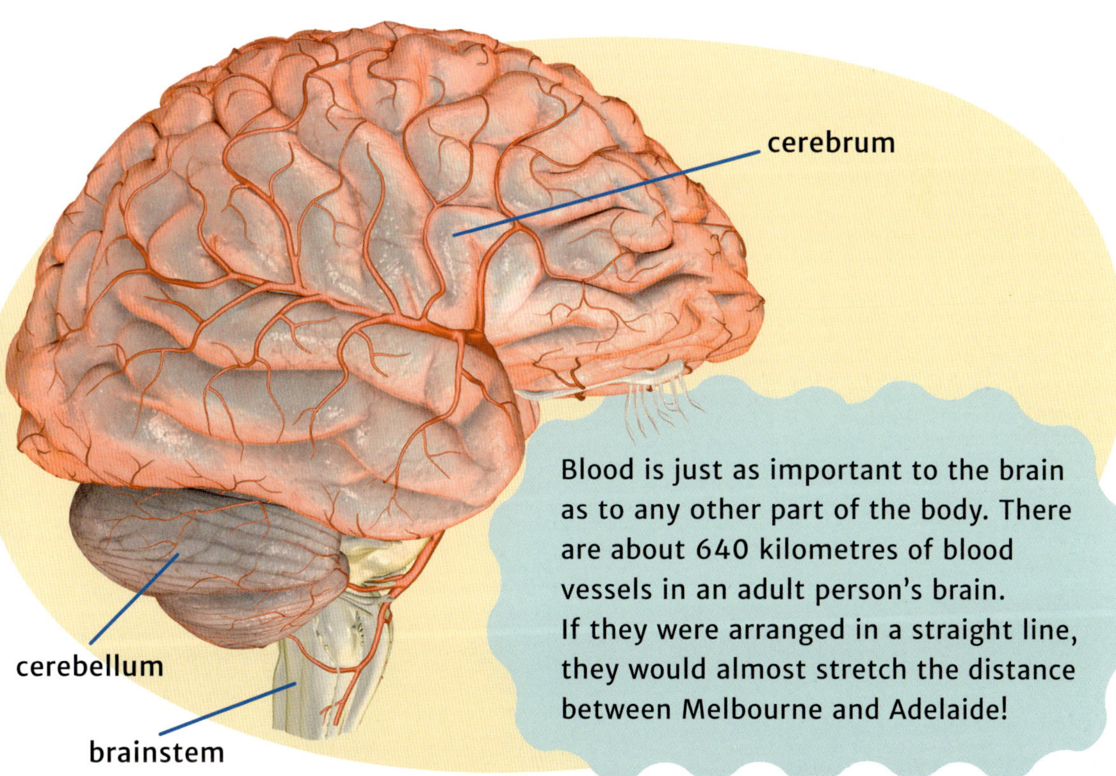

Blood is just as important to the brain as to any other part of the body. There are about 640 kilometres of blood vessels in an adult person's brain. If they were arranged in a straight line, they would almost stretch the distance between Melbourne and Adelaide!

The Cerebrum

The largest section of the brain, the cerebrum, is where people do their thinking. The cerebrum makes a vital contribution to people's ability to understand and use words, as well as to learn and to create memories. The cerebrum is also responsible for much of people's behaviour. This includes the movements they **consciously** choose to do, such as walking or picking up a pencil.

The outer layer of the cerebrum is known as the cerebral cortex. Much of the work done by the cerebrum takes place in this layer. The surface of the cerebral cortex is full of bumps and grooves. These bumps and grooves serve an important purpose – their shape allows more brain tissue to fit inside the cranium.

This model cerebrum shows how much folded tissue squeezes together to make the brain.

The cerebral cortex includes four main parts, known as **lobes**:

- the frontal lobe
- the parietal (pronounced *puh-rye-uh-tal*) lobe
- the temporal (pronounced *temp-or-al*) lobe
- the occipital (pronounced *ok-sip-it-al*) lobe.

Each lobe has its own special functions, but they constantly work together in a range of complex ways.

The Lobes of the Cerebrum

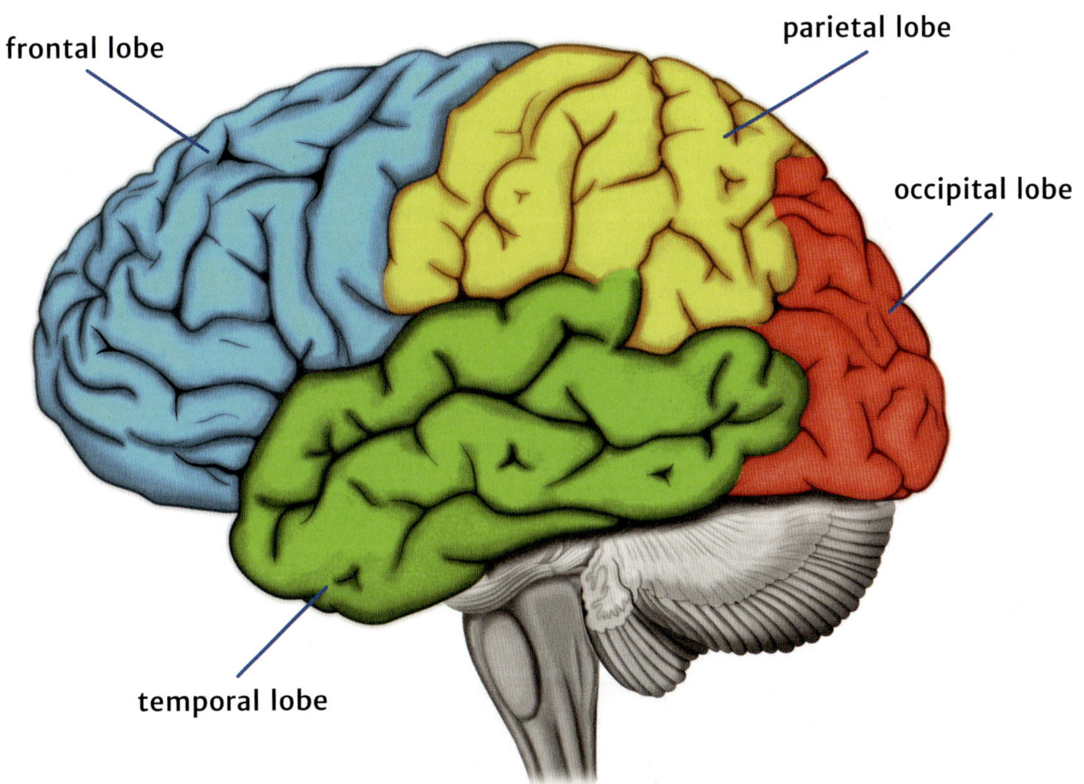

The Frontal Lobe

The frontal lobe of the cerebrum is the biggest lobe in the human brain. It is found at the front of the brain, near the forehead. The frontal lobe has two sides: the left frontal lobe and the right frontal lobe.

Each side of the frontal lobe affects the muscles and functions on the opposite side of the body. Therefore, the right frontal lobe controls the muscles on the left side of the body, and the left frontal lobe plays a similar role for the right side of the body. This is the same for the other lobes in the brain, too.

When a person does different movements with each hand, they are using both sides of their frontal lobe.

The frontal lobe has many important functions. One of its best-known functions is the role it plays in speech. It allows people to turn their thoughts into spoken words. The frontal lobe also controls many muscle movements, including those involved in walking and running.

People use their frontal lobe to walk and run.

The frontal lobe lets people tell each other what they're thinking.

The frontal lobe is very important for thinking, including the skills of problem solving and planning. It gives people the ability to focus their attention on particular tasks, or on things that are happening around them.

The frontal lobe also allows people to control their emotions, and to form lasting memories. Many of the **traits** that make up people's personalities are formed in the frontal lobe. These traits could include their sense of humour, and whether they are creative, friendly or quiet.

The American railway worker Phineas Gage showed a personality change after an injury to his frontal lobe.

The frontal lobe helps people focus on difficult tasks and solve problems.

This lobe plays an important role in the way people relate to each other, too. It allows people to feel **empathy**, so that they can understand how other people might be feeling. It also enables people to react **appropriately** to the feelings of others.

Empathy and reacting calmly are critical in caring jobs, like nursing.

The Parietal Lobe

The parietal lobe is found behind the frontal lobe, on the left and right sides of the head. It is particularly important for the five senses – sight, smell, taste, touch and hearing. Through these senses, people can perceive the world around them. This includes being aware of where the different parts of their body are, and how they are moving.

The parietal lobe allows people to perceive the world around them and where their body is.

Parts of the parietal lobe make it possible for people to sense temperature, and to feel physical pain. Although no one enjoys the feeling of pain, it is an important signal that something is wrong with the body. Feeling pain can prevent people from behaving in ways that would cause them harm.

Furthermore, the parietal lobe plays a major role in the ability to communicate. It helps people to understand the words they hear, and to recognise other people's voices.

Pain signals from the parietal lobe let people know when to rest an injury or seek help.

The Temporal Lobe

The temporal lobe is found in the lower part of the brain, near the ears, on the left and right side of the head. Like the parietal lobe, this part of the brain is important for hearing. It, too, contributes to people's ability to understand speech and recognise sounds. The temporal lobe also contributes to the enjoyment of music. It allows people to be aware of the volume of music, as well as its **pitch** and **rhythm**.

The temporal lobe helps to give meaning to sounds, like speech and music.

Additionally, the temporal lobe makes an important contribution to the sense of sight by allowing people to **interpret** the things they see. For example, it helps people recognise places or objects, including other people's faces.

Like other parts of the cerebrum, the temporal lobe plays a vital role in helping people learn and remember things. This occurs mostly in an area called the "hippocampus", where long-lasting memories are made.

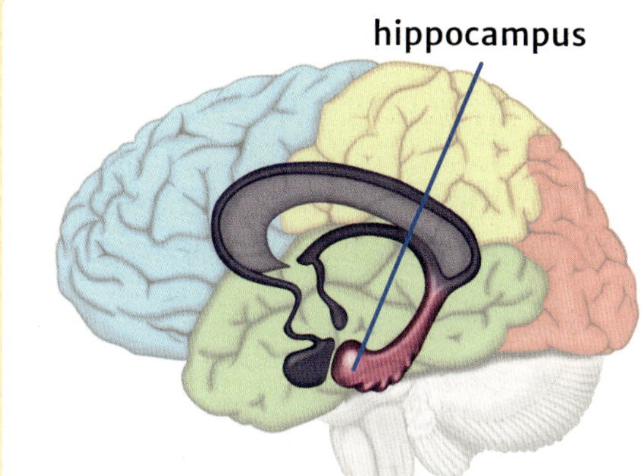

The hippocampus sits inside the temporal lobe.

People need the temporal lobe to watch media and play games, because it lets them interpret what they are seeing.

The Occipital Lobe

The occipital lobe is found at the back of the head. It is particularly important to a person's sense of sight. A part of the occipital lobe, the primary visual cortex, is connected to the eyes.

In the occipital lobe, information received from the eyes allows people to identify objects and to detect movement. It also helps people to judge if the things they see are nearby or far away. These abilities are very important for playing sport. For example, when people play tennis, they need to judge how near or far away the ball is, so that they can move into position and hit the ball over the net.

The occipital lobe allows people to see how close or far away objects are, and then interact with them.

The occipital lobe lets people see colours and tell the difference between them. It contributes to the ability to tell the difference between shapes, too.

The information received by the occipital lobe is sent to other parts of the brain. Working together, these parts of the brain help people to understand what they are seeing.

When children learn about shapes and colours, they are using their occipital lobe.

The Brainstem

The brainstem is attached to the spinal cord. It connects the cerebrum to the cerebellum and the spinal cord. Many vital functions in the body are controlled by the brainstem. Most of these are things that seem to happen automatically. These include the heartbeat, breathing and sleep. Like parts of the cerebrum, the brainstem is also involved in physical sensations, such as pain, hunger and thirst.

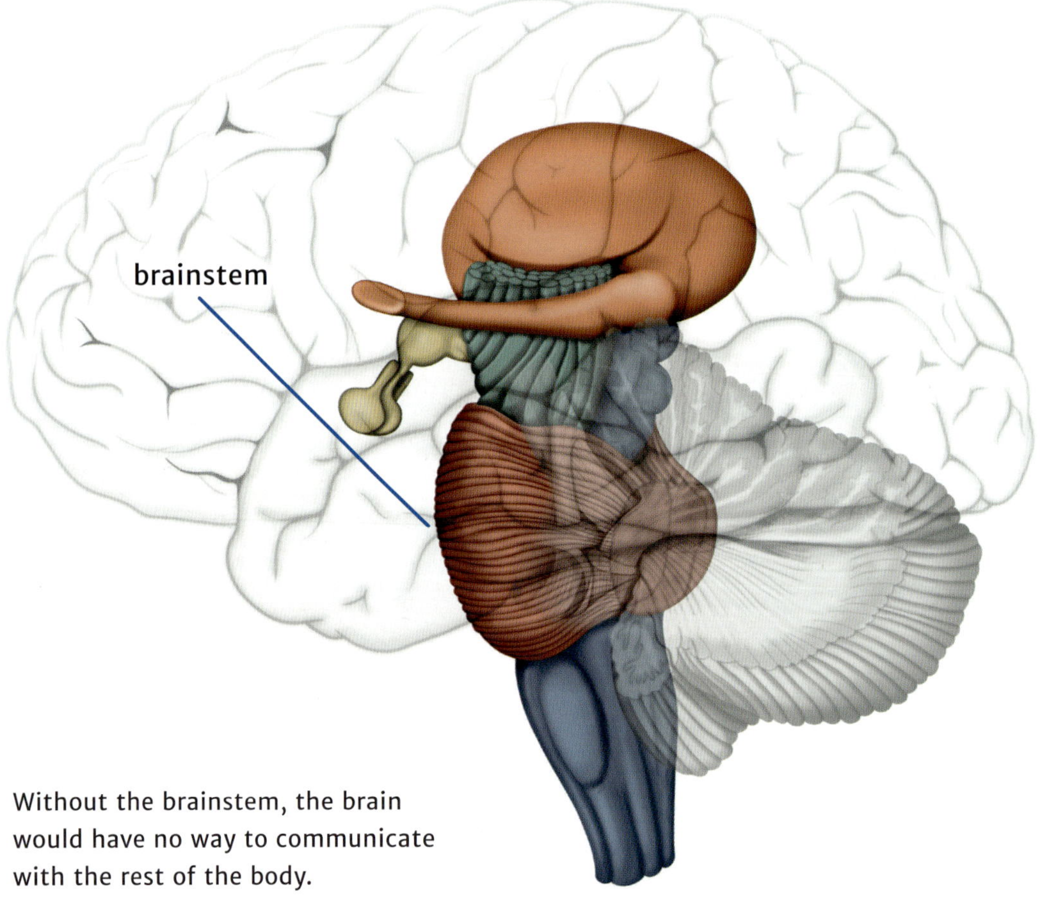

Without the brainstem, the brain would have no way to communicate with the rest of the body.

An important part of the brainstem is the medulla oblongata (pronounced *muh-duh-luh ob-long-gaa-tuh*). This is the lower area of the brainstem, which connects to the spinal cord. The medulla oblongata plays a significant role in some of the vital systems that keep people alive. For example, it controls the respiratory (breathing) system. It ensures that the **cardiovascular** system is working well, too, so that the heart can do its job of pumping blood through the body.

When people sleep, the brainstem helps their body deeply relax and keep breathing regularly.

The Cerebellum

The cerebellum is located at the back of the brain, between the brainstem and the cerebrum. Compared to other parts of the brain, the cerebellum is quite small, but it is involved in a wide range of vital functions. For example, it controls people's balance and "fine motor movement", meaning movement that requires precision. It also plays a role in speech, because it controls the muscles that people use when they talk.

Painting often requires fine motor movement.

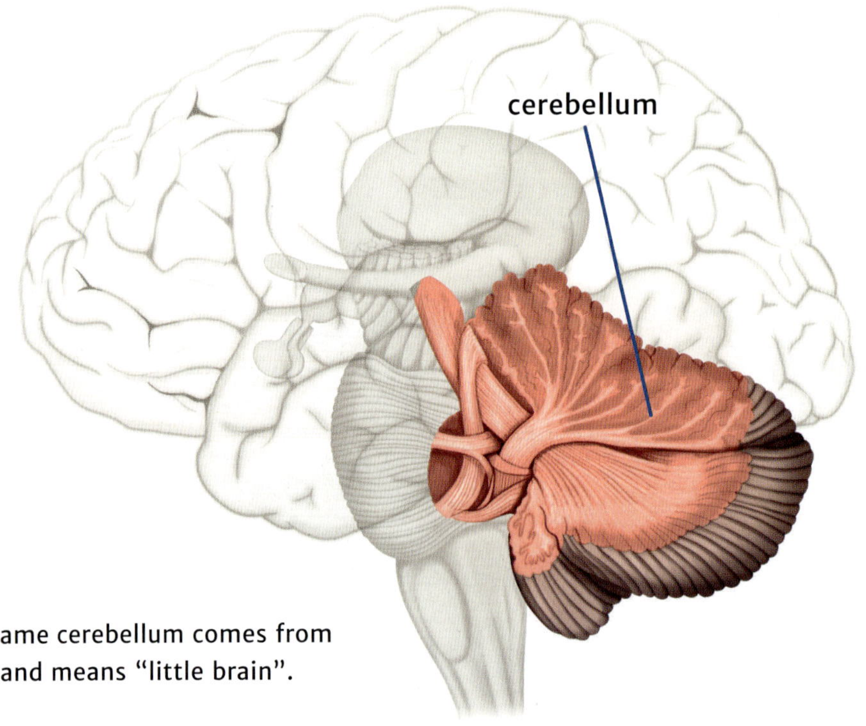

The name cerebellum comes from Latin and means "little brain".

Like most other parts of the brain, the cerebellum contributes to other important functions, including learning, especially when a person is learning activities that involve movement, such as dancing or riding a bike. It also plays a role in decision making, and in feeling emotions.

Sports stars can use their cerebellum to predict where a ball will go long before it gets there.

For the cerebellum, practice really does make perfect. The more you do an action, like riding a bike or catching a ball, the more the cerebellum will remember and let you do the same action again, often without thinking about it.

The human brain is truly remarkable. It allows people to understand and react to the world around them, and it allows the human body to function. Scientists are still learning about the functions carried out by the different parts of the brain, and about how these parts work together.

In the years to come, new discoveries about the brain will continue to help people understand themselves and this incredibly complex organ.

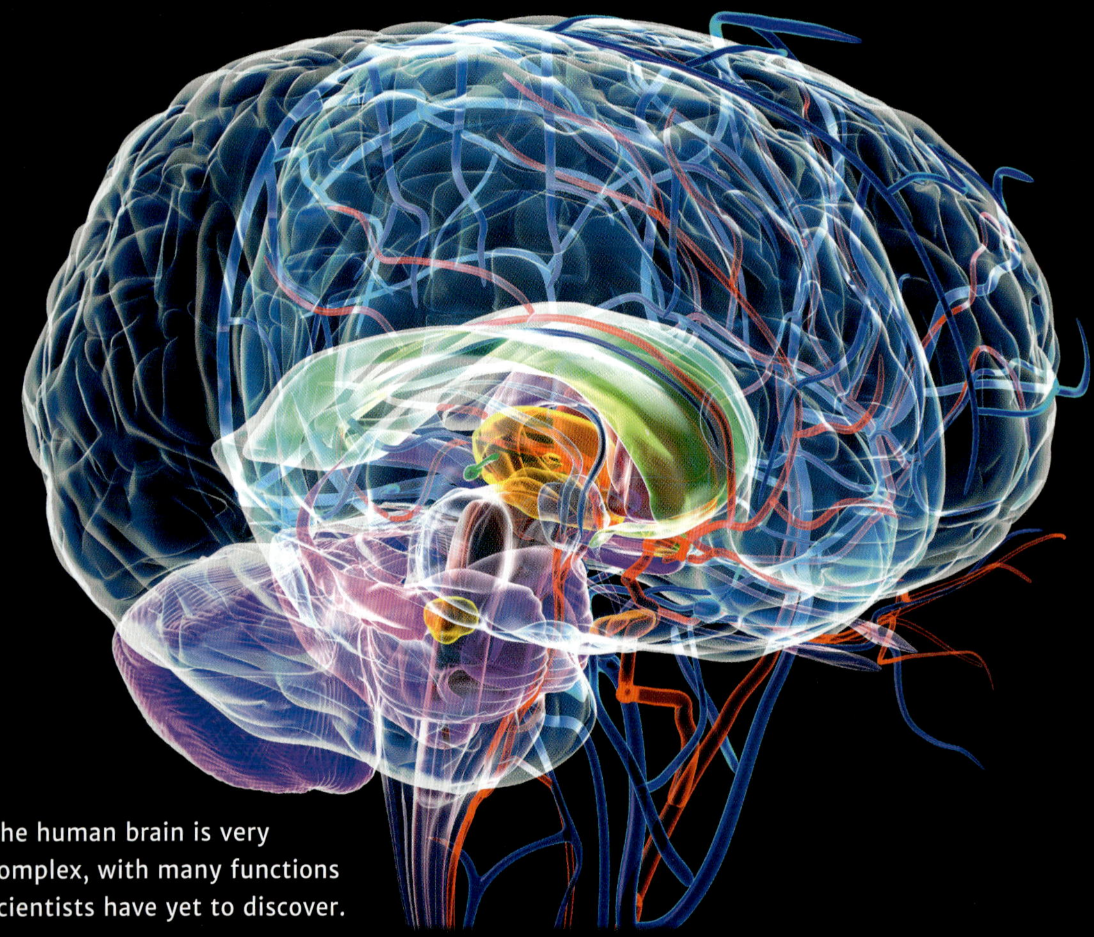

The human brain is very complex, with many functions scientists have yet to discover.

How Brain Cells Communicate

The brain is made up of billions of tiny **cells**. There are two important kinds of cells in the brain. The first kind are called neurons (pronounced *new-rons*). Neurons are like messengers. They are found in the brain and throughout the body in the nervous system. Every thought, movement or bodily function depends on communication between neurons. Most people's brains have about 86 **billion** neurons in total. A piece of brain tissue the size of a grain of sand can contain about 100 000 neurons.

Neurons produce electrical and chemical signals that pass information from one area of the brain to another. Neurons also pass information between the brain and the rest of the nervous system at great speed. They can do this because of the unique features that connect them with each other.

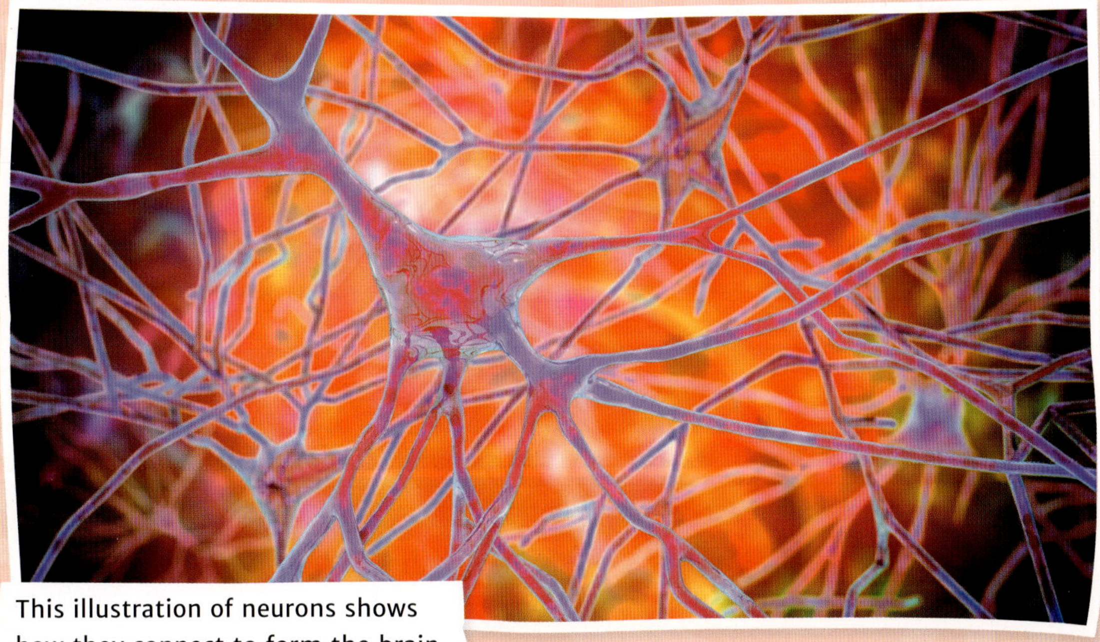

This illustration of neurons shows how they connect to form the brain.

There are many types of neurons, but they all tend to have some basic features in common. In general, neurons are made up of three main parts. These are a cell body and two parts extending from the cell body: an axon and several dendrites.

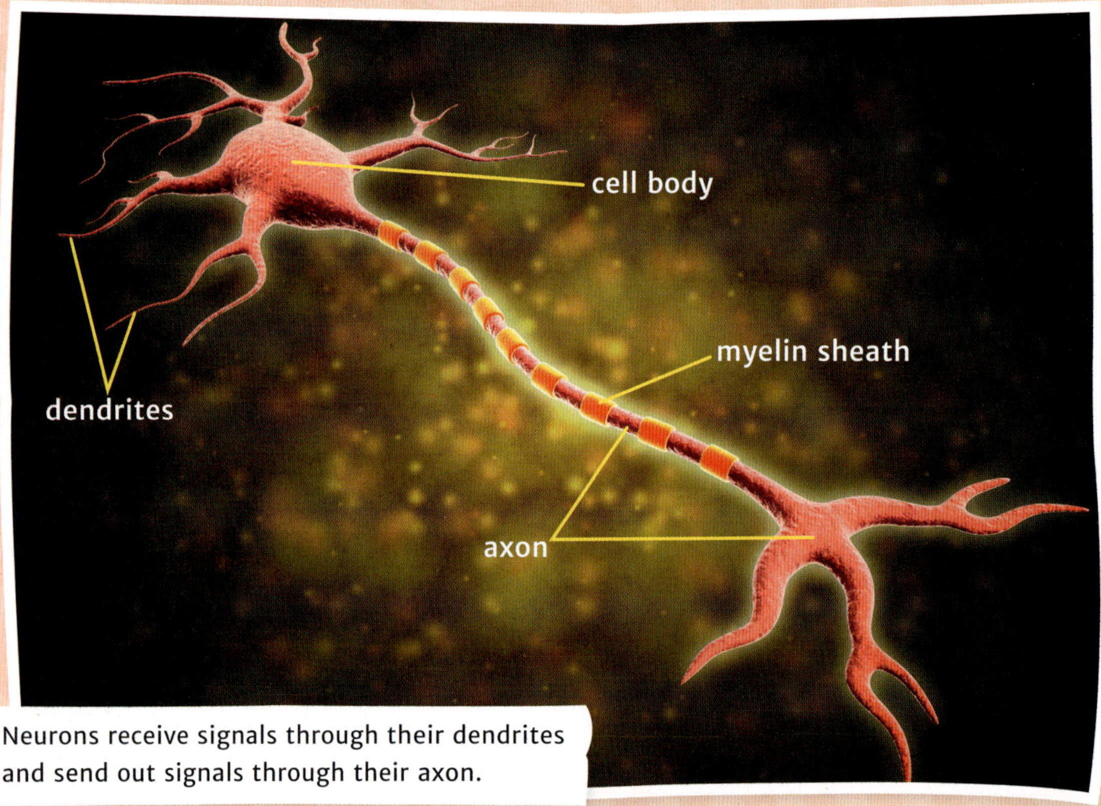

Neurons receive signals through their dendrites and send out signals through their axon.

The cell body is the main part of the neuron. It controls everything that happens in the cell. It also receives signals from other cells. If the signal it receives is strong enough, it passes it on to other neurons. Each neuron can send hundreds of signals each second.

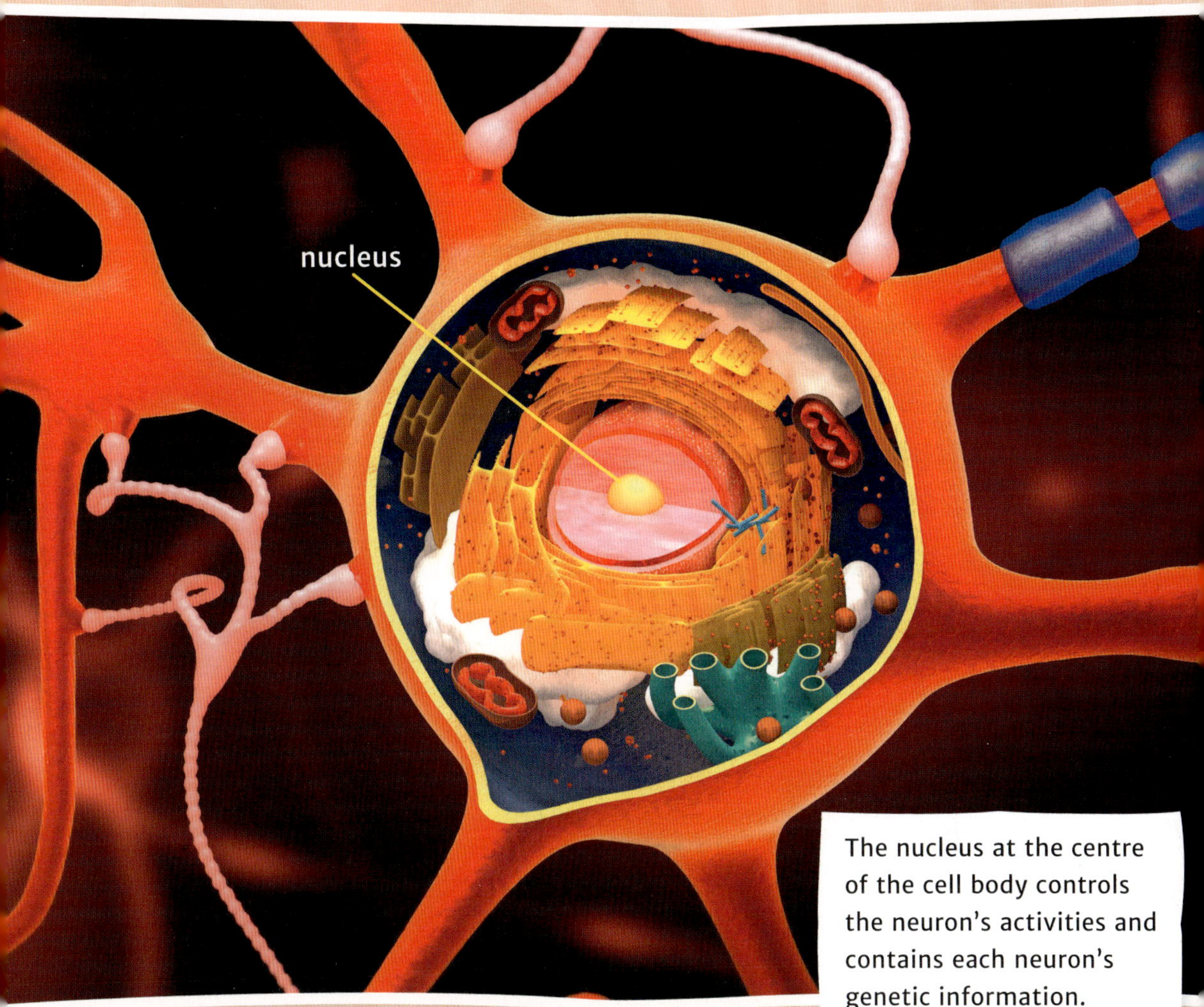

The nucleus at the centre of the cell body controls the neuron's activities and contains each neuron's genetic information.

Signals can travel from neuron to neuron at speeds of up to 430 kilometres per hour – faster than a Formula 1 racing car!

The dendrites are the parts of a neuron that receive signals from other neurons. They carry these signals into the cell body. Dendrites look like the branches of a tree. They reach out to connect with other neurons.

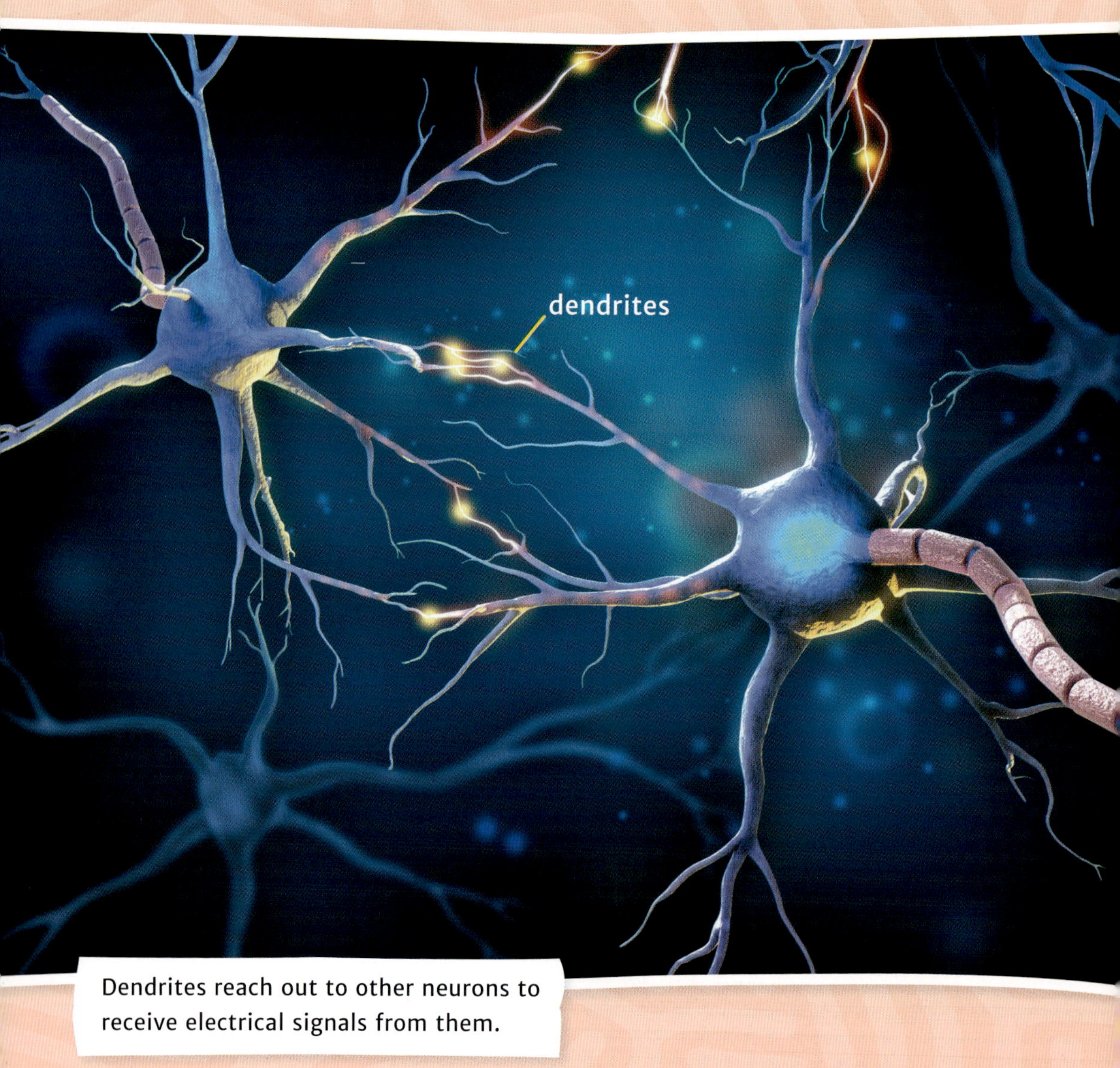

Dendrites reach out to other neurons to receive electrical signals from them.

The axon carries the nerve signal away from the cell body and towards the next cell. Axons are also known as nerve **fibres**. A neuron's axon looks like a long tail emerging from the cell body. The axons of many neurons are covered in a type of sleeve, known as a myelin (pronounced *mye-eh-lin*) sheath. The myelin sheath protects the neuron and helps the signals travel quickly and smoothly through it to the next cell.

Axons can be very long. The longest axons in the human body are more than a metre long. These are the axons of the sciatic nerve, which runs from the lower end of the spine down each leg and into the feet.

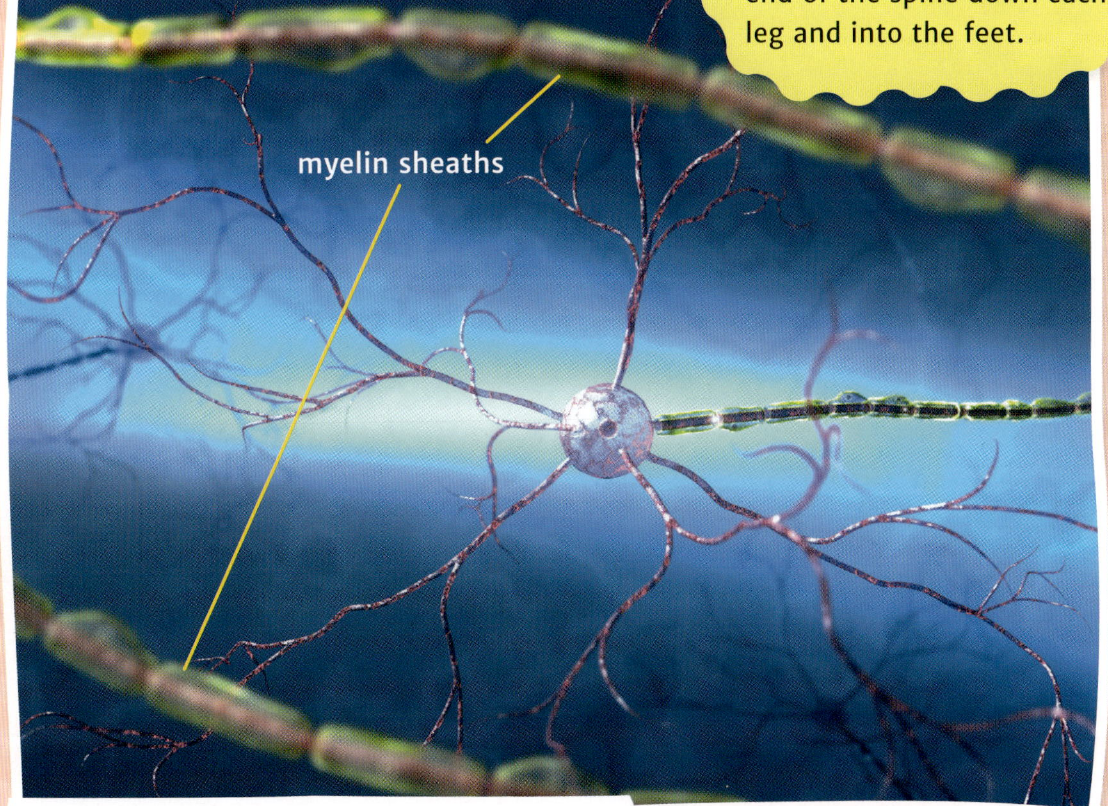

A myelin sheath is made of fatty substances that wrap around the axon fibres to protect them.

Neurons communicate with each other by sending chemical and electrical signals across a tiny space between them, called a synapse (pronounced *sye-naps*). These signals are sent from the axons of neurons to the dendrites of other neurons. They arrive at a special area of the dendrite called the **receptor site**. From there, the signals pass into the dendrite and through the cell body. Each neuron can send and receive up to 1000 signals per second.

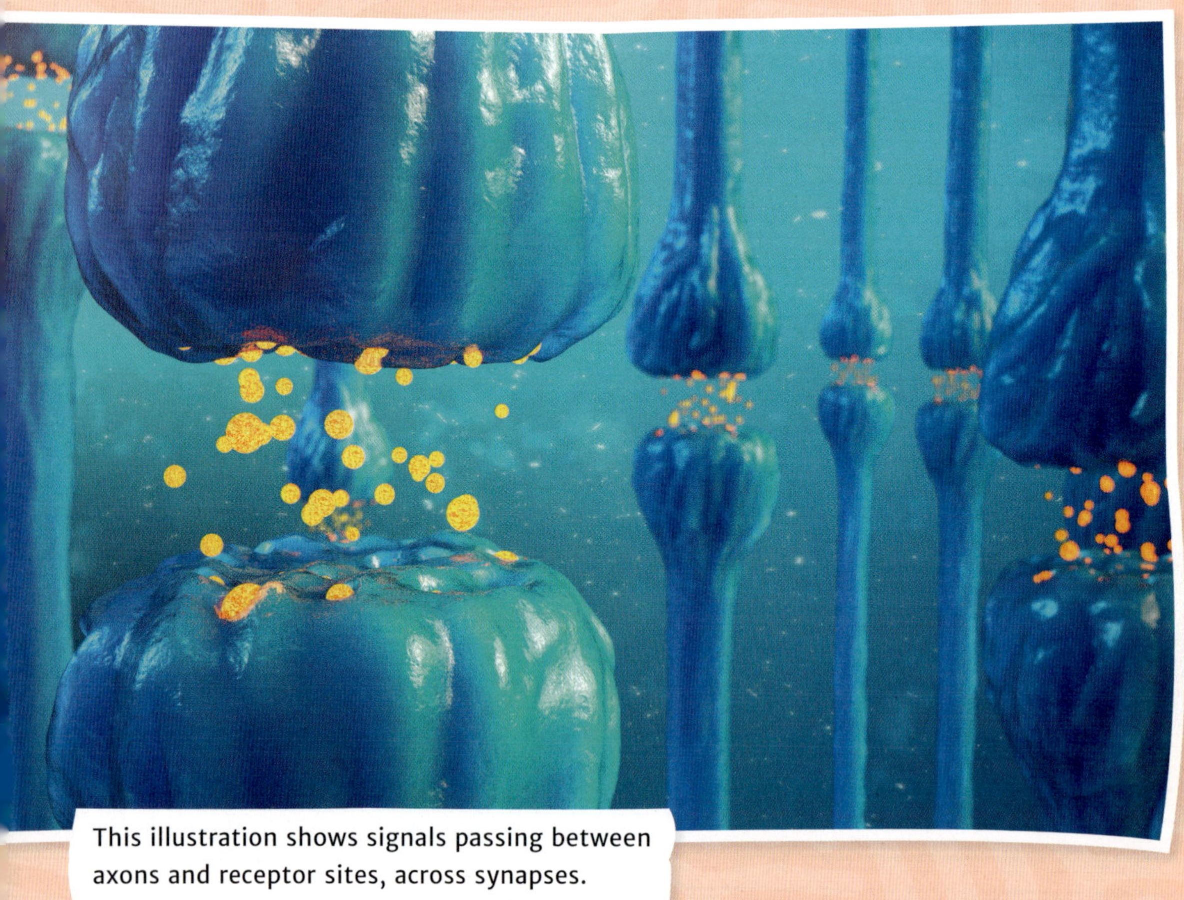

This illustration shows signals passing between axons and receptor sites, across synapses.

Although there are billions of neurons in the brain, neurons are not the only important cells there. The other important cells include glial (pronounced *glye-al*) cells. These protect and **nourish** the neurons and remove waste products to keep the neurons healthy. There are a similar number of glial cells in the brain as neurons.

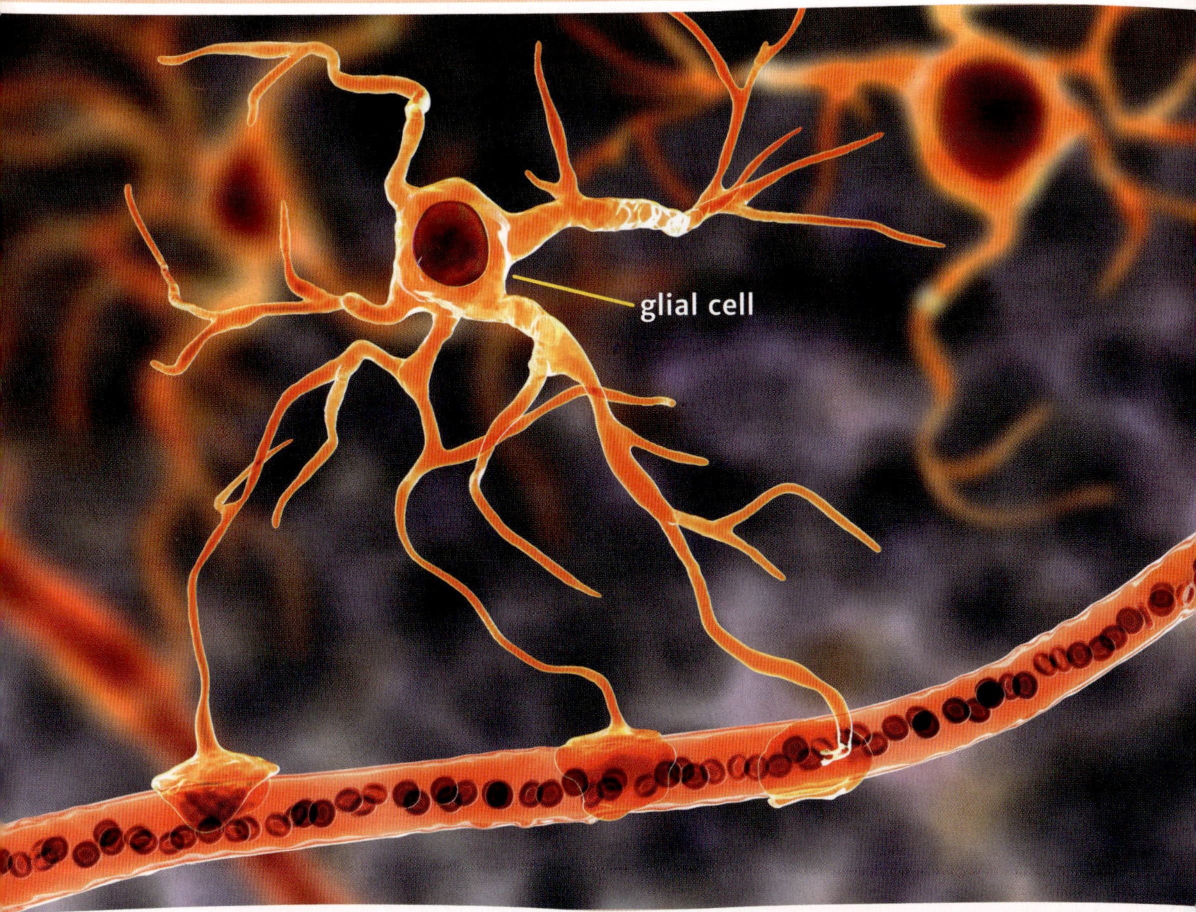

Glial cells are the cleaners of the brain, sticking to neurons and keeping them healthy.

The communication between neurons is vital to the functioning of the human body's most complex organ, the brain. This communication affects everything that people experience in their lives. Without it, people would not even be aware that they exist. They would not be able to learn, or to share the special experiences that make life so exciting.

The fun and meaningful experiences people have together are all possible because of the human brain.

Glossary

appropriately (*adverb*) suitably, correctly

billion (*number*) one thousand million

cardiovascular (*adjective*) to do with the heart and blood vessels

cells (*noun*) the tiny building blocks that all plants and animals are made from

consciously (*adverb*) with awareness of a thought or action

empathy (*noun*) the ability to understand how other people are feeling

fibres (*noun*) thread-like strands

functions (*noun*) important processes in the body

interpret (*verb*) to transform information into a form that can be understood

lobes (*noun*) larger parts of an organ

nerves (*noun*) threads that carry messages between the brain and the body

nourish (*verb*) to give something what it needs to grow and stay healthy

organs (*noun*) parts of the body, like the heart or the brain

pitch (*noun*) how high or low a musical note is compared to other notes

receptor site (*noun*) an area of a cell that can receive signals from other cells

rhythm (*noun*) the repeated pattern of short and long sounds in music

sensations (*noun*) feelings experienced when something touches or affects the body

spinal cord (*noun*) a column of nerves inside the backbone that connects all parts of the body to the brain

tissue (*noun*) a group of cells that work together to perform a function in the body

traits (*noun*) qualities or characteristics

Index